# Design your photovoltaic system

# Solar Panels 101

# 1st. Edition

Learn how to install and design your own solar panel system

Power your home, business, boat, RV, ranch and some applications.

## Samuel Edison

consult a licensed professional before attempting any techniques outlined in this book.

By reading this document, the reader agrees that under no circumstances is the author responsible for any losses, direct or indirect, which are incurred as a result of the use of information contained within this document, including, but not limited to, —errors, omissions, or inaccuracies.

# Dedicatory

Dedicated to my brothers...

What a bunch of assholes

# TABLE OF CONTENTS

# INTRODUCTION

When people think about alternative or renewable energy, the first image that comes to mind is often large blue or black solar panels on rooftops or portable highway signs that have a small panel attached. These panels, also known as photovoltaic modules (or PV modules), convert sunlight into electricity, and they have been the backbone of renewable energy for decades. The Photovoltaic Effect (how sunlight is converted into electrical energy) was discovered over a hundred years ago! Yet widespread implementation of this technology has been very gradual. Only in very recent years has photovoltaics gained wide popularity as an alternative way to produce electricity.

We are hearing much about small gadgets like calculators,watches, and small toy cars are being run by solar power. There are remote cell sites and other remote telecommunication mini-exchanges that are supported by it. So what is behind this technology? It comes from tiny solar cells combined together to produce the required electric power you need.

What is a solar cell then? It is also known as *"photovoltaic cell"* and it is a device that converts solar energy into electricity by the photovoltaic effect. These photovoltaic cells are assembled in arrays to form a solar module or photovoltaic arrays. Solar power derive its power from the sun by using solar panel. A small solar panel can be used to power up small devises like electronic calculators.

By combining a group of photo-voltaic arrays, it generates renewable electricity which you can substitute as a source of electric power to Radio equipments in the remote areas where no commercial power line is available. Photo-voltaic electricity is increasingly becoming popular in deploying in a grid-tied electrical systems. The definition from Wikipedia Encyclopedia of a grid-tied electrical systems is ,it is a semi-autonomous electrical generation system which links to the mains to feed excess capacity back to the local mains electrical grid. When insufficient electricity is generated, or the batteries are not fully charged, electricity drawn from the mains grid can make up the short fall.

Solar cells are primarily manufactured in the following countries: Japan, China, Germany, Taiwan and USA. Its life span is about twenty years. Just like when you want to build a house, you either hire an architect to draw up a plan for you or buy a model house with complete specifications. The same is true with building a solar power panel system. You need an expert professional to do it for you or buy a standard solar power system model.

Solar technology has improved a lot and the manufacturing process has been re-engineered many times. Hence, the end product keeps on becoming more attractive to the global community. It is also worth mentioning that we should continue doing advance research on how we can improve its performance.

Walk on a carpet with your socks on then touch a door knob, everyone knows the outcome. A nice shock to wake you up and a pretty blue spark. The friction of your movement across the carpet allows your socks to collect a charge. Your body then becomes a conduit for the charged electrons to flow. The door knob is the grounding rod that wants the charged electrons you have collected. As soon as you reached for the knob you became a pathway for the low of electrons.

Solar cells work on the same principle. The materials of the solar cell, mono or polycrystalline composites, have the ability to stop photon light waves. The action of stopping the motion of the light wave acts like the friction between your socks and the carpet. The photons give up there energy to the crystalline atoms of the solar cell. Now the solar cell is charged by the Photo Voltaic process. We need a way to move the charge now.

Until we create a pathway and storage to collect or use the energy. The solar cell will not collect more energy. Molded to the cells are electrodes, a pathway For the charged electrons to follow. Wired to a battery and inverter or a load. The most important part of the system. Like yourself being the pathway for the charge to transfer to the door knob. If you did not reach for the knob no electricity would have flowed. The door knob is the battery or load that needs the charge you have collected.

The cells in your panel are connected in series collectively to produce more output for load applications. The more cells or panels you have in series, the more wattage is produced. During the day when there is enough sunlight the panels produce enough electromotive force to run appliances. At night the batteries, that have stored an abundance of the energy produced by the panel, takes over.

In 1958 the first PV modules were launched into space to power satellites. Even today, solar power is the primary source of energy at the International Space Station. On Earth as well, PV has traditionally been used in areas where there is no practical source of electrical power but there is abundant sunshine. Solar panels are often used for remote applications: like powering cabins, RVs, boats and small electronics when grid service is not available. Recently, "grid-tie" solar electric systems have started gaining momentum as a cost-effective way to incorporate solar electricity into our everyday lives. Now we can take advantage of available solar energy while still enjoying the safety net of the utility grid.

In very basic terms, a solar panel (PV module) is a device that will produce a flow of electricity under sunlight. This electricity can be used to charge batteries and, with the aid of an inverter, it can power normal household electrical devices, or "loads". PV modules can also be used in systems without batteries in grid-tie systems.

Most PV modules are framed in aluminum, topped with tempered glass, and sealed by a waterproof backing. Sandwiched between the glass and backing layers are the photo-reactive cells themselves, often made of silicon. On the back of the module is a junction box that may or may not have two cables coming out of it. If the junction box has no cables, it can be opened to access the electrical terminals where wires can be attached to conduct the generated electricity away from the module. If there are cables already in place, the junction box is usually sealed and not user-accessible. Sealed junction boxes are more common.

There are lots of ways to make use of solar electricity. One of the simplest is to charge small electronic devices, like cell phones and music players, with lightweight, portable PV modules. These small battery-charging solar panels are even being integrated into backpacks and clothing for maximum convenience. These panels can be used individually or wired together to form a solar array.

For larger electrical loads, there are two main types of systems for providing electrical power to homes, cabins and offices, etc: stand-alone battery based systems (also called 'off-grid' systems) and grid-tied systems (also known as utility-interactive). You will want to decide which system is best for your needs by reading more about both.

There are many reasons why solar panel installation is on the rise like never before. The root cause is of course the huge current concern surrounding climate change and the devastation that could reek on the world in the worryingly near future.

With more and more scientists warning of the impending doom likely to be brought about by continued failure to address this issue, people are finally attaching real value to those measures that can have an instant and noticeable impact on greenhouse gas emissions. Of all the possible measures one could take, few are more immediate or significant than solar panel installation, as it instantly removes the households dependence on fossil fuels as a means of providing electricity.

While it is nice to be nice, we should not overstate the role that these pure environmental factors play in the decision by ordinary folk to switch to renewable means. The real incentive is that presented by governments keen for people to invest in solar panel installation. The UK government, for example, actually pay people for any surplus energy that their panels generate, that they are able to feed into the local grid.

It is this, combined with the great ongoing savings brought about by solar panel installation, that is motivating people to make the change. What is more is that these patterns are acknowledged by the property market that now appears to be adding a premium to any building that has a product installed, further fuelling the trend. All this contributes to an exciting scenario in which people from all social backgrounds and demographics are doing their art in the fight against climate change.

Do you know what is **DIY** stands for? It's **Do It Yourself**. Well, most people now know what a solar panel is since many residential properties use them to conserve energy and to lower the cost of their electricity bills. That is not all, solar power panels absorb heat and energy from the sunlight which can be used to run electricity in your house and it is a clean energy.

There are lots of manufacturers who have introduced do-it-yourself solar panels. All you have to do is buy the materials you need to make the panels on your own for your house. There is also kits wherein there are instructions on how to do it yourself.

If you try to think about it; why buy solar panels if you can do it yourself? You can buy a book for more ideas to learn how to do it. There are also ready made panels where in all you have to do is install it. If other people can make and/or install solar power panels, why can not you?

It is just a matter of determination to make your own solar energy panels. You can buy some manuals on how to do it yourself. Why not use your own determination and skills in making one. The point of DIY solar panels is to save money. So you have to expect trial and error in order for you to learn. No one is perfect and there will most likely be some bumps along the way.

As what you have always heard, experience is the best teacher and that is true. You can gather any information regarding solar power panels but it is best if you do it yourself. Although you can see that there is lots of DIY solar panels on the market today.

Would you not be proud if you could make your solar energy panels? DIY is the cheapest way. You can save cash for labor and installation. All you have to do is purchase the material you need to make the panels.

Buying DIY kits is also one way you can have your own idea where you can learn a lot regarding solar energy. The DIY panel was introduce to the market in order to give the people some information that everyone should know and learn how to do it by yourself.

The DIY solar panel is a great way to save energy and also save the environment where you live since it gives clean natural energy.

Electricity is one of the foundations of our world: our modern way of life would grind to a halt were there no electricity to run our myriad of gizmos and gadgets. As such, an entire field of engineering dedicated to keeping the juice flowing and finding ways of using that power to do all sorts of interesting things has developed. Most of us, however, are content just to know that if you plug your TV into the wall outlet and press the power button, you can watch your favorite diversions to your heart's content.

How important is electricity? Without it, the monitor with which you are viewing this book right now would not work, nor would the CPU which decodes this web page and tells your monitor what to display. Given the importance of computing and the internet or email in the worlds of business, finance, science, and daily life, this alone is a rather important reason to ensure that the juice keeps flowing.

Besides its role in the recent computer and internet revolution, electricity plays a vital role in modern industry, whether to power industrial robots or to harness the raw power of huge electrical currents to melt iron in steel foundries.

So, just what is electricity? The basic definition is quite simple to understand: electricity is a flow of electric charge. Charge is a fundamental quantity in physics (i.e., it cannot be described in terms of yet more basic concepts) which all of the fundamental subatomic particles which make up our world carry (neutrons are not fundamental particles but are made from charged particles whose charges happen to cancel).

The amp, short for ampere (named after eighteenth century French physicist Andre-Marie Ampere) is the basic unit of electric current. Electric current is quite analogous to a flow of water: the amount of "stuff" that goes past a point in a certain time is the current at that point. In the case of water, or any other fluid, engineers use a term called volume flow rate, which is the volume of said fluid that passes through a certain cross-sectional area per unit of time (e.g., X cubic meters per second). The definition of electric current is similar, except that the "stuff" being measured is the amount of electric charge passing through an area such as a point on a wire.

Voltage, on the other hand, is a somewhat trickier concept to put into familiar terms. What we call "voltage" is actually an informal reference to electric potential difference (no wonder we find it easier to say "voltage" instead!). The closest analogy to a flow of water is to the pressure difference in a pipe; absent any greater momentum in the opposite direction, water will flow from the area of greatest pressure to the area of least pressure. While pressure creates a "potential" in the case of water, it is a difference in charge which creates this potential in the case of electricity: a current is simply nature's way of trying to even out an imbalance of charge by sending negative charges to an area where there is a relative deficiency of negative charge (i.e. a more positively charged region), or vice versa.

A "volt," then, is a unit of electric potential difference. The simplest way to think of the number of volts is as the amount of desire electrons in the current have to get across the distance separating the differently charged regions. That is, the "voltage," or potential difference, across a gap of a given distance varies with the amount of the difference in charge from one side of the gap to the other. If one side of the gap has five units of positive charge and the other side five units of negative charge, the voltage will be five times greater across the gap than if there were one unit of positive or negative charge on each respective side.

The watt (named after Scottish inventor and engineer James Watt), finally, is a unit of power, just like the horsepower is a unit of power. Since power is the same quantity in the eyes of physics whether it comes in the form of electrical power or mechanical power or any other form of power, nothing about the watt is particular to electricity.

# WHAT ARE SOLAR PANELS?

The sun is the primary source of energy on Earth and sunlight can be converted directly into electricity using solar panels. Electricity has become indispensable in life. It powers the machines that most us use daily.

So, what are solar panels? What if you can create your own?

## A .The Components

A solar panel is usually manufactured from six (6) components namely the PV (photovoltaic) cell or solar cell which generates the electricity, the glass which covers and protects the solar cells, the frame which provides rigidity, the backsheet where the solar cells are laid, the junction box where the wirings are enclosed and connected, and the encapsulant which serves as adhesives.

Since most people does not have access to equipment in manufacturing solar panels, it is important to note and understand those six components in order for anyone to be able to plan the materials needed to create a do-it-yourself or home-made solar panel.

The materials needed on how to make a solar panel must be available for purchase locally or online and should not exceed the cost of a brand new solar panel or does not take a long time to build.

- ***Photo- voltaic Cell***

The first thing to consider when building your own solar panel is the solar cell.

Photovoltaic (PV) cell or solar cell converts visible light into electricity. One (1) solar cell however is not enough to produce a usable amount of electricity much like the microbot in Baymax (Hero 6) which only becomes useful when combined as a group. This basic unit generates a DC (direct current) voltage of 0.5 to 1 volt and although this is reasonable, the voltage is still too small for most applications. To produce a useful DC voltage, the solar cells are connected in series and then encapsulated in modules making the solar panel. If one cell generates 0.5 volt and is connected to another cell in series, those two cells should then be able to produce 1 volt and they can then be called a module. A typical module usually consists of 28 to 36 cells in series. A 28-cell module should be able to produce roughly 14 volts (28 x 0.5 = 14VDC) which is enough to charge a 12V battery or power 12V devices.

Connecting two or more solar cells require that you have a basic understanding of series and parallel connection which is similar to connecting batteries to make up a battery storage system.

There are two most common solar cells that can be bought in the market; a monocrystalline cell and a polycrystalline cell. These two can have the same size, 156mm x 156mm, but the main difference would be efficiency. It is important to purchase additional cells to serve as backup in case you fail on some of the cells i.e. bad solder, broken cell, scratched, etc.

Monocrystalline solar cells are usually black and octagonal in shape. This type of solar cell is made of the highest and purest grade silicon which makes them expensive. But they are the most efficient of all types of solar cells and are almost always the choice of solar contractors when space is an important factor to consider in achieving the power they want to attain based on their solar system design.

Polycrystalline PV Cells are characterized by their bluish color and rectangular shape. These cells are manufactured in a much simpler process which lowers the purity of the silicon content and also lowers the efficiency of the end product.

Generally, monocrystalline cells are more efficient than polycrystalline cells but this does not mean that monocrystalline cells perform and outputs more power than polycrystalline cells. Solar cell efficiency has something to do with the size of the cells and every solar panel or cells have an efficiency rating based on standard tests when they were manufactured. This rating is usually in percentage and the common values range from 15% to 20%.

## ▪ *Glass*

The glass protects the PV cells while allowing optimal sunlight to pass through. These are usually made of anti-reflective materials. Tempered glass is the choice of material nowadays even for unknown and new manufacturers although there are still those who utilize flat plate glass on their solar panels. Tempered glass are created by chemical or thermal means and is many times stronger than plate glass making it more expensive to produce but the price of manufacturing them today is reasonable and cost-effective. Flat plate glass creates sharp and long shards when broken as opposed to tempered glass which shatters safely in small pieces upon impact, that is why they also call it safety glass. It should be noted here that most amorphous solar panels uses flat plate glass because of the way the panel is constructed.

Tempered glass is what manufacturers use in mass producing their solar panels. In our DIY project, we suggest to use Plexiglas also called acrylic glass which is safer than the regular normal glass from your local hardware store. It is a bit expensive than regular glass but is weather resistant and does not break easily. The Plexiglas can also be screwed or glued easily to the frame.

- ***Frame***

A frame is usually made of anodized aluminum which provides structure and rigidity to the solar module. These aluminum frames are also designed to be compatible with most solar mounting systems and groundin equipment for easy and safe installation on a roof or on the ground.

The frame in a factory-built solar panel is usually the aluminum part where all four sides of the solar panel sheet are inserted. Think of it as a skeletal rectangular frame. The solar panel sheet by the way is composed of the other 4 components and are layered and laminated in the following order from top to bottom;

the tempered glass, top encapsulant, the solar cells, bottom encapsulant, then the backsheet. In our DIY solar panel, we will be using a wooden frame and the end-result would be something analogous to a picture frame where the picture is the solar cells glued to a non-conductive board, the glass for the Plexiglas top cover, and the wooden part as the frame and backsheet.

- **Backsheet**

The backsheet is the layer of plastic film on the back surface of the module. This is the only layer protecting the module from unsafe DC voltage. The main function of the backsheet is to insulate and protect the handler from shock and provide the safest, efficient, and dependable electrical conductivity possible.

The backsheet will be a wooden plywood where the frame will be screwed on top and on the sides. It should be noted here that a perforated hardboard (Pegboard) will be used to place and align the PV Cells and this Pegboard will sit on top of the wooden backsheet and fitted inside the wooden frame.

- ***Junction Box***

The junction box is where the terminal wires and bypass diodes are located and concealed. The terminal wires are basically the positive and negative wires based on the series connections of the PV Cells and can be connected to another solar panel, a charge controller, a battery system, or to an inverter, depending on the system design. The by-pass diode is a protective mechanism that prevent power from getting back to the solar panel when it is not producing electricity as in the case when it is night time.

There are junction boxes designed for factory-built solar panels that are now available to purchase online especially from China. If you are not pressed for time, you can order online and wait for the delivery otherwise you can just purchase a regular electrical junction box from your local hardware store. The purpose of the junction box is to protect the terminals (positive and negative terminals) from water, dust, and other elements. This is also where the two wires (red for positive and black for negative) will be coming from. The other end of these two wires can also be protected by using a PV accessory called MC4 which can also be purchased online together with the PV junction box.

- ***Encapsulant***

Encapsulant sheets prevent water and dirt from infiltrating the solar modules and serve as shock-absorbers that protect the PV cells. They have this adhesive bonding capability to the glass, the PV cells, and the backsheet similar to a glue but stronger. Encapsulants are usually made of Ethylene-Vinyl Acetate or EVA and are applied using lamination machines and processes. Solar panel manufacturers use a vacuum and a large oven to properly seal and cure the EVA sheet onto the solar panels. Most of us do not have the capability to do this but many still have tried and failed while others had varying levels of success.

Encapsulants are thin plastic sheets that are usually laminated on the top and bottom parts of the solar cell sheet. The bottom encapsulant is the layer on top of the backsheet where the solar cells are actually placed and supported.

# TYPES OF SOLAR PANELS

To build solar panels, you will certainly need to know the type of solar panels you actually want to power your home. As the technology is persistently improving and many new types of solar cells are invented, it is essential to understand the main difference between them.

✓ *Mono crystalline(known as Single Crystalline)*

Mono crystalline solar panels are basically considered as the most efficient ones. The main difference from other solar panels is that these are made from one large chunk of silicon crystal. They are among the oldest and most reliable silicon cell technologies.

The process of making these large silicon crystals is a very energy demanding process, which adds up to the final solar system cost. Certainly, they are considered the most efficient, able to produce electricity at 15-18% efficiency, but not necessarily the best opportunity for home owners.

One of the arguments, why people should purchase mono crystalline solar panels is because there is not a large or, in fact, a small space on the roof to install them. With mono crystalline solar panels you can be sure, that you use the available space on the roof in the most efficient way possible.

The main difference in appearance is that they are black in color and they are rounded in form (cells).

Also these panels can last from 25 to 50 years on maximum, so they are a good long-term investment.

However, they are also fragile, so that some care is necessary. Therefore, a rigid frame is more than appropriate.

In short, mono crystalline panels are best for efficiency, performance and longevity. The negative - more costly than other types of solar panels.

✓ *Polycrystalline*

Polycrystalline solar cells, as the name indicates, are made up of multiple silicon crystals, not like mono crystalline cells. Usually, they look somewhat similar to a mosaic. In color, they are dark ocean blue.

In general, polycrystalline solar panels are among the cheapest and most widely found panels on the market today.

At slightly lower efficiency rates than mono crystalline solar panels, they still are able to produce electricity at about 12-14% efficiency. Moreover, they are produced with less energy wasted. That is why this technology is constantly evolving today.

Polycrystalline cells are a great alternative to mono crystalline cells, because they offer a slightly better cost-per-watt efficiency. Thus, many people prefer this type of solar technology today.

One thing must be remembered though - they are designed to work best at relatively cooler temperatures. It is beneficial to know that temperatures starting from 60°C and up can decrease the sunlight-electricity conversion ratio by more than 20 percent at these temperatures.

Ever wondered, why sand in the beach feels warmer than air on sunny days? Well, that is because sand is a better conductor than air. And the same principles apply to silicon cells, as sand contains silicon. So you need to watch the temperatures and control them for maximum electricity generation.

Nevertheless, polycrystalline solar cells are usually the best opportunity for a DIY home solar panels project.

### ✓ *Amorphous (thin-film)*

Amorphous solar cells are one of the newest types of solar cells. These are very versatile, as they can be used to produce electricity in ways crystalline technology would not be able to. Basically, the silicon atoms are not ordered in a crystal lattice like in crystalline cells.

With this technology, they are not growing; crystals, but the silicon is deposited in a very thin layer onto a backing substrate. Although the production process is complicated, they manage to produce amorphous cells with less energy. These panels are less time-consuming and expensive to make, so that they can be produced with better efficiency.

Another great advantage for amorphous cells is the ability to be flexible. This is possible because of the thin layers of silicon, that are applied. Today, there are already many creative uses of flexible amorphous cells available for different uses. For example, attachable to handbags, e-readers, clothes, etc.

However, amorphous panels have many drawbacks, especially in efficiency. They are only 5-6% efficient, which is not much for a potential solar system installation for your home. Likewise, the high impurity levels can cause drops in sunlight-electricity conversion ratio, when the solar panels begin to generate electricity.

Remember, these are the most common types of solar cell technologies, which have proven their reliability along the way, so they are a very good start to look into more depth, when considering to build your own solar panels. There are a lot of other emerging solar cell technologies, but have to be developed and put to test.

To name a few, there are group III-V technologies (used in aerospace) which are very expensive; string ribbon photovoltaic cells, which are evolving, offering, in some instances, higher efficiency levels than poly silicon and are also cheaper to make; BIVP - serving as both an electricity producing and building construction material; concentrator systems, which use lenses to gather sunlight into a concentrated form to increase solar cell efficiency; multijunction devices and others.

Plenty of great technologies to choose from, but the decision, which type of solar panels to go for, is all yours to make.

# FACTORS AFFECTING PHOTO-VOLTAIC CELLS

## ➤ *Inverter Efficiency*

When the solar PV system is catering to the needs of the AC loads an inverter is needed. As things stand, in real world nothing is 100% efficient. Although inverters come with wide ranging efficiencies but typically affordable solar inverters are between 80% to 90% efficient.

## ➤ *Temperature*

High temperature can severely reduce the solar panel's production of power. Higher temperature increases the conductivity of the semiconductor, charges become balanced within the material, reducing the magnitude of the electric field, inhibiting the charge separation, which lowers the voltage across the cell. Depending on the location, heat can reduce the output by 10% to 25%.

In the built environment, there are a couple of ways to deal with high temperature. Install solar panels on a mounting system a few inches off the roof, this will help cool them by allowing air circulation. Use photovoltaic panels that are designed to be more efficient in hotter climates. Ensure that panels are constructed with light-colored materials, to reduce heat absorption. Inverters and combiners can be moved into the shaded area behind the array.

## ➢ *Shading*

Ideally solar panels should be located such that there will never be shadows on them because a shadow on even a small part of the panel can have a surprisingly large effect on the output. The cells within a panel are normally all wired in series and the shaded cells affect the current flow of the whole panel. But there can be situations where it cannot be avoided, and thus the effects of partial shading should be considered while planning. If the affected panel is wired in series (in a string) with other panels, then the output of all those panels will be affected by the partial shading of one panel. In such a situation, an obvious solution is to avoid wiring panels in series if possible.

> ### *Dirt and Dust*

Dirt and dust can accumulate on the solar module surface, blocking some of the sunlight and reducing output.

> ### *Battery Efficiency*

Whenever backup is required batteries are needed for charge storage. Lead acid batteries are most commonly used. All batteries discharge less than what go into them; the efficiency depends on the battery design and quality of construction; some are certainly more efficient than others.

# ADVANTAGES OF SOLAR PANELS

A solar panel is a device that is used to absorb energy from the sun in order to generate heat or in many cases electricity. It is also referred to as a photovoltaic cell since it is made of many cells that are used to convert the light from the sun into electricity. The only raw material for these solar panels is the sun. it is made in such a way that the cells face the sun in order to enable maximum absorption of the sun rays. The greater the energy from the sun is, the more the electricity that is generated. Solar panels are used in many homesteads in the world due to their many advantages that are far more than disadvantages.

**The advantages are;**

- *No emission*

One very important advantage of using solar panels is that they do not emit any gases that are common in green houses. The panels do not emit any smoke, chemical or heavy metals that can be risk factors to human health. Solar panels are therefore environmental friendly when compared to burning of fossil fuels to generate energy. This is very important since carbon emissions are dangerous and avoiding their emission helps in safeguarding our present and future environment. Being environment friendly is important since the government is constantly coming up with ways to control global warming and the use of solar panels is a great way to start. The solar panels therefore maintain a clean setting and they leave the air fresh. More importantly they help in prevention of many cancer incidences. This is because some products from some sources of energy like nuclear energy have been said to cause cancer due to initiation of mutations in cells.

- **_Free energy_**

Use of solar panels ensures ongoing free energy for those who use it. This is mainly because the only cost incurred is that of installation. Once the installation has been done the energy is free since the panel does not require regular maintenance or fuel to run it. It also requires no raw materials for its operation. It works as long as there are sun rays which is an everyday thing in most parts of the world. In a world where equal distribution of resources is continuously being sought, this is very important since each and everyone has equal rights when it comes to use of solar energy. This is because the energy from the sun falls on all. This is a good way to maintain equality as compared with energy from fossil fuel which low income homesteads do not afford in many cases.

• *Power decentralization*

There is also the advantage in that, the use of solar panels enable the decentralization of power. This is very important since it is very cheap. This is mainly because when power is not decentralized, it has to be shared by all and is as a result transported to many areas. With this happening, there are very many costs that are incurred. These include; the wear and tear of vehicles, the air pollution among others. These costs are all incorporated in the electricity bills of individuals as the government does not cover the expenses. It is therefore more advantageous to use solar panels as a saving plan and to create a sense of fairness since those in power tend to take advantage and use their positions to embezzle funds. This is not fair on the citizens' part. This is because most of them struggle to make ends meet.

- *Operate off-grid system*

A solar panel can be operated off grid. This is a great advantage for those who live in very isolated areas or in rural regions. Off grid means that the house is not connected to the state's electricity grid. This has the advantage of low cost since installation may be very expensive for those living in isolated areas. These individuals have their power lines disconnected in many instances due to the fact that it is sometimes less affordable for many. Solar panels offer a solution for this since they do not require as much to be installed. However, those living in towns can also use the off-grid technique. An added advantage in this is that there are no rules governing whether or not one wants to operate off- grid or on-grid when it comes to use of solar panels. This however is an issue when using fossil fuel generated electricity.

- ***Generate job opportunity***

Solar panels generate job opportunities. This is of great importance since there is a very high rate of unemployment in the world today. These jobs are come about in the form of, manufacturing of the solar panels, research about further improvements, maintenance, development and cultural integration. With the continued presence of the sun, these jobs are guaranteed since there is ongoing improvements and modification of this device. Jobs like maintenance and installation do not require a long-term training and are therefore more advantageous for those who do not have many skills and are unemployed.

- ***Price control***

Use of solar energy is safe from price manipulations and politics. The fact that there are no raw materials that are solely controlled by monopolies ensures that there is no manipulation of prices as is the case with fossil fuels. With fossil fuels, the prices can rise as high as the monopolizing powers controlling them want. There is also less competitiveness with use of solar panels since there is no fight over such things as oil fields and other raw materials. Although the government has started addressing the issue of solar panels, there is little influence they can have in price manipulation. This is because no one controls the main raw material.

- ***Protection or preservation of environment***

There is also less environmental destruction with the use of a solar panel. This is because there are no cases of mining or extraction of raw materials that eventually lead to destruction of forests and water catchment areas. With the use of solar panels, there is less of this and therefore there are steady rainfalls that greatly boost production and consequently the national income of each and every country. Many countries face problems of famine due to destruction of forests to get fuel. This can be prevented by using solar panels.

- *Reliability*

There is an advantage of reliability in using solar panels. This is because there is ability to predict the amount of sun to expect each and every day. Therefore one is has guarantee of energy. The devices are also made in such a way that they can absorb sun rays even when there a few clouds and the sun rays are not very strong. The solar energy is also renewable. It can therefore be used on and on without getting depleted. Although solar energy cannot be used at night, it operates full force during the day which is of great importance. The energy can also be stored in form of batteries for use at night.

- *Noise free*

Everyone loves some peace and quiet. This is something you get when you use solar panels. This is because they are very silent. There is no noise that gives away the fact that the solar panel is there apart from the fact that you can see it. This is a good thing since it makes the environment peaceful compared to wind and water generated power supplies which have moveable parts that are quite loud and destruct the peace. Solar panels are therefore good for use for people living in estates where hoses are close to each other. This is because with silence, peace is maintained between the neighbors.

- *Minimal space require*

When installing solar panels, there is no large scale installation required. They therefore require very little space to install. This is very important when it comes to fast growing regions and towns. The installation will mainly involve a single cell to continually generate energy. Hence a homestead requires a single cell. There is therefore no congestion and a continued supply to the high demand of energy. This maintains a good image in a community since crowding may make the place less attractive which may prevent people from moving to the area since everyone wants to live somewhere they consider beautiful, for this reason, use of solar panels does not interfere with real estate sales.

- *Durability*

Solar panels are durable. This is because there are no moving parts in the device. This therefore reduces the chance of it being destructed. It is possible to use a solar panel for a very long period of time without having to purchase another, studies estimate that it can last for over ten years. Such a device is beneficial because it reduces the stress that comes about when a machine stops running because something became lose or worn out. There is also reduced maintenance cost since it is less prone to wear. This generally makes the device very easy to handle for a person with very little skills in handling a solar panel.

- ## *Its profitable*

Many companies that invest in solar energy get the advantage of higher profits. This is because they cut back on costs incurred in electricity and the rest of the profits are in most cases used to expand the business. This is very advantageous. Statistics show that the companies that use solar panels have greater returns compared to those that use other sources of energy. This may be due to the fact that electricity can be very expensive and may make these companies not afford allot of assets. This is especially evident in small or new companies. There is also an advantage that clients get when they get services from a company that uses clean energy. This is the fact that they can get access to government incentives that are made available to these companies.

- *Low tax charged*

Use of solar panels enables individuals and companies to enjoy the benefits of low taxes. This is because in most parts of the world, the taxes that are charged are about thirty percent less as compared to using other sources of energy. With all the taxes that one has to pay for every item purchased, this is a great opportunity to reduce spending on taxes. Since there is no monthly bill when using a solar panel, it makes it tax free. When using fossil fuel energy, this is no option since one has to pay their electricity on a monthly basis which in most cases is heavily taxed.

- *Low energy consumptions*

The size of solar panel required per meter to give maximum energy small. When there is full sun, one is able to get about one thousand watts per meter. This is equivalent to about 2900-watt hours per day.

However this depends on the area in which you are locates, the time of the year and the strength in which the sun rays reach the solar panel. For this reason, there are times where one gets more energy compared to others. However the energy gives the desired effect even at low intensity and is therefore still very reliable.

- *Safety*

It is highly unlikely to hear that someone got injured when using a solar panel. This is because there are few cases of electric shocks that are very frequent when using other sources of electricity. It is therefore safe to use solar panels for people. This creates fewer incidences of emergencies. However, careful measures should be taken as directed by the person that does the installation since there are instances where cables are left bare and could cause shock when touched. This is rare when the wiring is done correctly. Care should also be taken since the roof could be constantly emitting electricity.

- *Environment friendly*

Solar panels are not prone to destruction by harsh environmental conditions. For this reason, they are not easily destructed, this is important since the device is placed outside in order to absorb sun rays. The good thing about this is that it can be used by people who live in areas where the weather is up and down in most cases.

All these are great advantages that come with using solar panels. Solar panels can be used in any setting, whether in schools, homes or companies

# HOW TO PROPERLY INSTALL A SOLAR PANEL

Solar energy is the most economic and environment friendly way to heat, get electricity and water with just a little endeavor. It is the preferred way to get sufficient energy for the fulfillment of your daily needs. Solar energy has made the use of solar panels progressively more popular. Solar power system can actually gather the energy from the sun and convert it into renewable energy for your everyday use. This book will reveal  a few points you need to consider when installing solar panels.

## 1. Take Permission from the Local Zone Authorities

Prior to the installation of solar panels, make sure you have permission from the local authorities, as these are banned for domestic use in many cities due to aesthetic reasons. Contact the local zone authorities of your area to own a license prior to the installation.

## 2. Decide the Right Place

They are usually installed on rooftops in order to receive the direct and maximum amount of sunlight. It is vital to install the solar panels at a place where it is directly exposed to the sun in order to perform at optimum capacity. The position angles can easily be calculated relying upon the latitude on which they are being installed. Make sure that the structural integrity of the roof can support the hefty panels for your home.

## 3. Mounting of the Solar Panels on the Roof beam

The next step to installing the solar photovoltaic panels is mounting. Three main varieties of solar panel mounts available are pole mounts, roof-ground mount system, and flush mounts. Using these mounts you can either install these on the roof or affix them as a free standing unit. Generally roof-ground mount system is used to attach the panel to the roof and also provide support to the panel from the bottom. The roof-ground mounts also permits to adjust the system to produce the required power output. Make sure that mounting is placed at a distance of about 48 inches and located directly on top of the roof beam. Each must be mounted to the solar alignment and interrelated properly. Each mounted part must be checked carefully to make sure that they are safe and sound and leak-proof.

## 4. Anchor the Solar Panels to the Racking System

With the help of a pilot bit, drill a hole in the mounts to keep the rafters together and for preventing splitting. Tie up the base of the mount by using stainless steel lag bolts. The post of the mount should be fixed into the base. Keep in mind to place the roof metal flashing over each of the mounts to prevent the roof from leaking. Connect the solar racks made up of aluminum to the metal rails to complete the racking system. Now attach the panels to the racking system with the help of an adequate hardware. Be sure that both the panels and the racking system are appropriately beached in accordance with the local electrical codes.

## 5. Setting Up of the Junctions in the Electric Circuit

Finally, interconnect the panels by setting up all the junctions present on the back of each panel and fix the electric wires into their appropriate terminals. After interconnecting the electric wires, close all open junction boxes.

Your solar panel is installed now and ready for use. You can now switch on the lights and use your own energy source. Indeed, the solar technology will offer you several years of satisfactory and competent service.

# WHERE TO INSTALL YOUR SOLAR PANEL?

Solar panels offer a cost-efficient way to generate renewable and natural energy. Solar energy is a reliable energy source that saves you money on energy costs and helps protect our environment. Solar panel installation is the best way to generate solar energy and transform your house into a power generator. Solar electric system consists of photovoltaic cells that convert sunlight into a renewable and clean energy. If truth be told, solar panel installation is a valuable option for creating your own electricity for daily expenditure

## ❖ *Install the Panel in Direct Sunlight*

Prior to installing your solar electric system, it is very critical to identify the right and appropriate place of installation. It is very important for the solar electric systems to get maximum amount of sunlight in order to perform at optimum capacity.These electric systems are usually installed on the roof, building tops or stand-alone units to receive maximum sunlight. The solar panel installer will help you to identify the right position for the installation of photovoltaic cells so they can directly receive sunlight for maximum efficiency.

## ❖ Remove Unnecessary Objects

While selecting the right location for installing you photovoltaic unit, make sure you choose a location where trees, trim branches, foliage and other unnecessary objects will not obstruct the sunlight. As these objects may cause daytime shadows which will have a marked effect on the efficiency of your system.

## ❖ Check the Roof Support Capability

Another important consideration while installing your panel is to check the supporting capability of the roof area. Make sure that your roof is structurally sound and has enough capacity to support the solar electric systems.

## Mounting a Solar Electric System

After determining the right location for installing your panel, another important consideration is the mounting which is used to install photovoltaic cells. There is a variety of mounts available for solar electric (photovoltaic) panels including pole mounts, flush mounts or roof-ground mount. Your solar panel installer will help you out in opting for the right type of mounts.

Solar electric system installation will provide you the most promising renewable energy source that will definitely reduce your monthly energy expenditure. It is a lucrative and cost-effective way to generate nonpolluting, safe and clean solar energy for indoor heating and electricity generation. Selecting the right location for installing a solar electric panel is very critical. Placing solar photovoltaic cells in the right place will boost the efficiency of the solar power system. Getting the help of the professional solar power system contractor is a worthwhile decision in this regard. He will better guide you in determining the right location and choosing the appropriate type of your solar electric system.

# REASONS WHY YOU NEED A DIY SOLAR PANELS GUIDE

DIY solar panels has not to be the most difficult task on Earth, but sometimes doing it yourself can be a daunting process. Some people do not have the patience, some get stuck in a certain step, but others want to have somebody show them the whole process from start to end step by step. Here are some common reasons why you need a guide to build solar panels.

✓ *Value of Saving Time*

The most important reason for a DIY solar panels guide has to be the time savings you can get. Simply enough, when you get an instructional guide, all the information is laid out for you ready to use right away. All is gathered together. No searching in forums, blogs and other websites needed.

The time you can lose doing massive searches for information, how to build solar panels from scratch, is tremendous. Usually it goes like this: you find some information on how to start building it yourself; then you initiate all the DIY efforts, but soon hit a step, when you do not really understand, what to do in it. Then all the time-lasting searching occurs. It is not uncommon to search for weeks for the necessary information.

Contrariwise, with a guide you eliminate all that searching by simply following the outlined steps. It is very unlikely that you will get stuck somewhere, if the guide has all the necessary steps included to get the job done. It is your decision to make how valuable your time is.

### ✓ *Watch a Real Person Doing It*

It is essential to have someone show you exactly how to make a solar system from scratch. And this is the case where video proves to be the best. Most of us easily absorb information visually better than by reading or listening to audio.

With a comprehensive DIY solar panels guide you can expect to finish the work in no time, if you attentively follow the outlined steps.

## ✓ *Advice From a Seasoned Builder is Gold*

Who would you choose: a person who has built a solar panel a dozen times or a person who has done it one time?

A seasoned builder knows better the common mistakes you can make along the process. He can guide you through all the shortcuts and make life easier for you. It is always easier to do what you are told than invent it yourself.

## SOLAR CELLS MODULES

A photovoltaic module is a package of solar cells which are interconnected into an array. They are more commonly known as a solar panel and used in a wider application such as residential or commercial solar power generation.

Each PV module is limited in the amount of electricity that it can produce so many installations contain several modules connected creating a solar array. A typical photovoltaic installation contains several photovoltaic modules, a inverter, connecting wire, and batteries.

Solar panels use photons from sunlight to generate electricity through the photoelectric effect. The solar cells are typically crystalline wafers built out of silicon.

**To use the PV cell in practical applications-** *They must be connected to each other in a* *system*

Protected from mechanical damage during transport, installation, and usage. Plexiglas is usually used to protect the cells in an enclosure. Solar cells are incredibly brittle. They break easily in transport and under extreme weather conditions.

Protected from moisture - The tabbing wire and cells need to be protected from moisture so that conductance is optimized. Corroded tabbing wire and solar cells lead to lower efficiency.

Electrical connections are made in series to produce optimal output. Blocking diodes are used to avoid overheating due to constant disconnects that happen as a result of limited sunlight exposure. It is also important to make sure that the system receives enough ventilation so the system does not overheat.

New technology called concentrators are now being developed which use lenses and reflective materials to concentrate sunlight into beams which are directed at the solar cells. This increases the output of the overall solar panel. Rather than installing more cells, it makes better usage of the current cells already in place.

# SOLAR ARRAY- Determining PV Array Maximum System Voltage!

Photovoltaic solar panels come in different wattage sizes and are designed to supply energy to your home. Generally, solar panels are classified by their rated output power which is given in **watts**. This wattage rating is the amount of power that a single solar panel can produce in one peak hour of sunlight. One of the biggest technical challenges to overcome with all photovoltaic installations, regardless of configuration, is the correct sizing of the system to meet the demands of the household.

The size of the photovoltaic system required varies from home to home as each homes energy usage and energy efficiency will be different. But determining the optimum number of panels and total wattage of your solar system on requires knowledge of your household usage and some simple maths. To help you overcome some of these challenges, this book will put together an easy-to-follow, step-by-step guide that will assist you to easily size your photovoltaic system.

### ❖ *Determine the Suns Peak Hours Available Per Day*

Solar panels are typically sold by the peak watt. When the sun is at its strongest or peak intensity usually at midday on a clear day, it produces about 1000 watts per m2 of solar radiation directly onto the Earth's surface. One hour of maximum, or 100% sunshine received by a solar panel equals one equivalent full sun hour. So if a solar panel is rated at say 100 Wp (peak watts) it would supply 100 watts of peak power at the brightest part of the day. If the average peak sun hours for a particular location is given as 4.5 hours, this means then that our solar panel will provide 450 watt-hours a day of peak electricity.

Obviously the sun shines longer than 4.5 hours a day. Climate data given for a particular location on the

Earth's surface would give the solar intensity data in terms of peak sun hours, so the suns intensity from sun rise to peak hours and back down to sunset throughout the day will be a percentage of the peak hours and therefore the power output from a photovoltaic cell will also be a percentage of the maximum during these times. For example, early in the morning a 100W solar panel may only be producing 25 watts, then midday it produces the full 100 watts, and in the afternoon only 25 or 30 watts again.

### ❖ Determine Your Energy Needs In Terms Of Watts Per Hour

To determine the required overall power rating of a photovoltaic solar system required to power a home, the electrical energy needs in terms of watts per hour should first be evaluated. To work out your homes power requirements, you need to do some homework first. Everyone's power consumption is different so by listing and adding together the appliances, lights and TV's with their hourly power requirements in terms of watts you will arrive at the total watt-hours per day you need.

The final power rating of the solar system can then be calculated and sized, based on the portion of the homes electrical energy consumption to be supplied by the system. So for example, a system that is required to supply 100% solar electricity would be twice the size of a system designed to supply only 50% of the consumption. Then a photovoltaic system can be sized to provide part or all of your electrical consumption.

❖ *Optimize Your Power Demands and Usage*

The ability of a photovoltaic solar system to produce free electrical energy is not unlimited. It is limited by the number of hours a day the sun shines and it is limited by the physical area available to install the solar panels. Accidentally leaving on a light bulb on during the day can easily consume and waste unnecessary amounts of energy. Saving and reducing your energy needs by using energy-efficient light bulbs and appliances not only saves you money but done correctly can reduce the final size and cost of your new solar photovoltaic system.

Solar systems are designed for a certain amount of energy consumption, and if the home exceeds the planned limits this additional energy will need to come from the utility grid costing you money. An energy-efficient home reduces the number of solar panels required making the installation of the system cheaper, less complicated and reducing its payback period so lower your power consumption and reduce your power needs as much as possible.

❖ *Determine The Type of Solar Panels You Wish To Use*

There are many hundreds of different size solar panels available to choose from ranging from 50 watts to 250 watts per panel at 12, 24 or 48 volts and all with their own set of advantages and disadvantages. The number and type of solar panels required to capture enough solar energy to support your electrical consumption plays an important role in the design, sizing, operating voltage and cost of your solar photovoltaic system.

A typical solar panel is made up of a grid of individual solar cells. There are different types of solar cells to consider. Monocrystalline silicon solar panels are the most efficient at converting the suns solar energy to free electricity, but they are also the most expensive. Polycrystalline silicon panels are slightly less efficient than monocrystalline, but they tend to be cheaper since they are cheaper to produce. Thin film solar panels are the least efficient, but they are also the cheapest. Thin film solar panels are uniquely versatile as the silicon film is thin and flexible. Shop around the market to find the best panels that suit your needs.

## ❖ *Size Your Solar Array*

To estimate the size of your solar array, you will need to divide the previously calculated total watt-hours by the peak sunlight hours you should get the total wattage of solar panels that you will need and then add a little extra to make up for cloudy days. This gives us the total number of solar panels we need to generate a given amount of Watt-hours (or kWh) for our home in our given location. For example if we need a 1000 watt system, that is 10 x 100 watt panels or 5 x 200 watt panels.

Since the solar panels will be used to supply the home directly with free solar electricity or to charge batteries, it is necessary to decide what the nominal DC voltage of the system will be. Depending upon the required battery storage and inverter sizing, the configuration of the solar panels may be connected in a series configuration, a parallel configuration or both. If you want year-round reliability, it is best to use the lowest DC voltage and power rating possible to reduce breakdowns and to keep our solar electrical system running effortlessly and economically for years to come. The peak power rating of the solar panel you will be using can be found in the manufacturers specifications.

Sizing a solar array is not as difficult as you may think, but there are two factors to consider first to make your life easier. 1), What is the average amount of sun hours per day in your local area (which can be found from the city hall or library) and 2), what is the daily power consumption of your electrical loads. The sunlight is the sunlight and there is not a lot you can do to increase it, but lowering the electrical demand of your home can save you a lot of money in the long term, as well as reducing the size of your solar array.

But there are electrical loads that are not cost-effective to power using solar energy as their consumption would be more than the solar array could supply. Any load that requires electricity to generate heat such as water heating, space heating, cooking, air conditioning, etc. all these devices should be powered by other means.

## REASONS TO INSTALL A SOLAR ARRAY

With all the talk of alternative energy on the news lately, you might begin to wonder how you can be a part of it. There are many ways to become involved from hybrid cars to geothermal heat and cooling. One of the best ways, however, is solar. Really,here are some excellent reasons for installing solar panels on your home.

*1) Save Money* - Electricity rates have been going up for quite some time now and they are not expected to reverse this trend. Really,some experts believe that electricity rates could skyrocket over the next few years. With solar, you insulate yourself from these problems. Once you install a complete system you never have to pay an electric bill again. That is a great feeling. Even if you install a system that only supports a portion of your needs, you are cutting down your electric bill each month.

**2) *Power Outages Do Not Happen* - *If*** you have ever lived through a multi-day power outage you know how very inconvenient this is to put it mildly. With a complete solar array, you will always have power even if your neighbors do not(depending on how you have your array configured). You cut your dependency to the power company. In fact, this can feel liberating which might be yet another reason for solar!

**3) *Increase Your Home's Value* -** Solar is a very desirable item in the real estate market. Of course, if you intend to move very soon installing a solar system does not make sense. However, if you are going to spend many years in your home and ultimately desire to sell, it can increase your selling price. A research revealed that a solar system that saves $1,000 per year on electric bills increases the home's value by $20,000.

**4) Protect the Environment** - Solar power is renewable and does depend on fossil fuels. For this reason, solar is a clean technology. By investing in a solar array you are helping to make the earth a cleaner place.

**5) Technology is Cool** - This is not an oft cited benefit but it is a reality. A solar array is a great conversation starter as people have a interest in it. Plus, if you have children or grandchildren it is a great live science experiment. Teaching about the various aspects of the system can lead into many interesting and educational conversations. By talking to people and your children you are helping to spread the word about solar power and helping the earth and others in the process!

Of course, solar power is a complicated subject with many aspects to consider. Take the first step and call an installer to see what it would take to install a small, starter solar system. Your pocketbook and the earth will be glad you did!

# DIAGRAM OF PV CELL

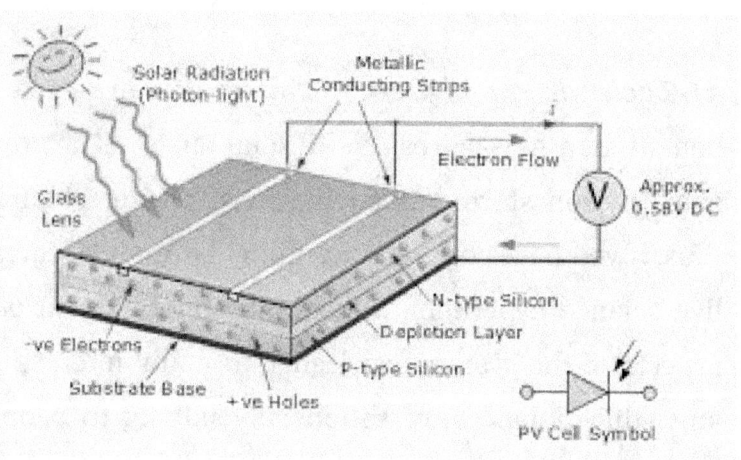

# THE DIFFERENCE BETWEEN OFF--GRID AND GRID --TIED SYSTEMS

These are two basic kind of systems you need to know before furthering yourself with any kind of plans to build a project using renewable energy. When building a solar power system, building an off grid or a grid tied system will be the two choices you will have to consider when you pursue your project.

the basic concept behind an off grid system is purely what it means. It is not being connected to any commercial power and having total independence. In order to do this, you need to calculate the total consumption of what your house or cottage consumes in watt hrs. This is also how the power company monitors your power consumption too. Now, according to the total amount of watt hrs you consume in that period of time is going to be relative to the sizing of your battery system you are going to need to store that power. Once you have stored power an inverter will be needed to convert from DC power to AC power.

A grid tied system, is a system where you will still use commercial power for your appliances. You can also incorporate a battery system, if you wanted to operate an off grid system also. But mainly what this does is if any power that is generated is greater that that power consumed commercially, can be sold back to your power company. this can be done with a special type of inverter which is directly connected to you meter box.

Both of these systems can be powered by solar panels that can built by anyone. Possibilities are within reach with a step-by-step guide which will show you how to do this. With a little research comes development.

# GRID-TIED SOLAR SYSTEMS

The majority of the homes in some part of the world are connected to the electricity grid, so it is most common for people wanting to include solar panels in their power system to want to integrate them into their grid system. A system like this can have batteries, but very often homeowners will forgo the batteries and just have their system grid-tied only.

### Grid-tied solar systems components.

**Solar Panel Collection:-** Once you have assessed how much energy you wish to produce and how much of your monthly consumption you wish to offset, you will know how many panels to have in your collection.

**Inverter:-** Your solar panels will be creating DC which you will need to convert to AC in order that it can be used to run appliances in your home.

**Breakers and Disconnect:-** This will ensure that your system is completely safe and that any power surges will be dealt with without damaging your overall system.

### Other neccessities aside components

**Interconnection Agreement:-** This is a legal document that you will receive from the utility company outlining in full details of your grid-tied system. This document will tell you how to tie your particular system into the grid and when you can expect them to inspect the system.

**Net Metering:-** This will allow you to sell your excess power back into the grid.

## The Advantages Of A Grid -Tied System

- These systems are easier to install than an off-grid system
- You do not need batteries
- You can sell your excess power to the grid and offset your costs
- You can often get rebates through incentives for installing PV systems

*The only disadvantage of this system is that it is only available to people who live on-grid.*

## Equipment For Grid-Tied Solar Systems

- Central inverters
- Micro-Inverters
- Power Meter
- Solar panels
- DC disconnect
- AC breaker panel
- Kilowatt-hour meter
- Utility disconnect
- Electrical wiring

All of these components function together in an intricate style to collect and distribute clean, renewable energy. It starts using the solar panels that collect sunlight and convert it into an electrical present. The electricity that these panels produce is direct existing (DC) power, but since your residence as well as the grid function on alternating existing (AC) power, the raw solar energy requirements to be converted.

# WHAT IS AN OFF-GRID SOLAR SYSTEM

An off-the-grid solar system is a system of solar panels that make a house or a household entirely or close to self-sufficient in terms of energy production. Although it is very hard to build a truly autonomous home, the energy system can be autonomous at the very least. And that is very viable for rural houses or cabins. Furthermore, an off- grid solar system can be used merely to power the conditioning of the house to keep the attic nice and cold during the summer months.

An full off- grid solar system usually consists of: solar panels (a solar array) that generate power; a PV combiner box - a box which protects the system from short circuits; charge controllers, which ensure that the batteries do not go overdrive; and a current inverter, which turns the solar panels' DC power to usable AC electricity. An electrical generator (wind or fuel powered) is optional, but in many times necessary, as unpredictable weather patterns can disrupt the power supply.

Solar panel efficiency is increasing day by day as the technologies progress. Ambitious green living enthusiasts predict that solar energy will be cheaper than grid electricity in the future,at least in the more sunny regions. However, note that solar panels do have an expiry date, and you should always check the projected "life expectancy" of the panels you opt to purchase. This will ensure the viability of your project. In sunny, off- grid solar systems can vastly decrease the power costs, as the initial payment for getting the grid to a remote location can be quite high.

Australia has the second cheapest electricity across the globe, but still there are numerous advantages to using an off- grid solar system even if you do not save that much money. First of all, you are doing planet earth a favour by using power that is acquired from renewable resources. Secondly, if you have a generator, you are more protected against power failures than other inhabitants of the same area. And lastly, an autonomous power supply is an excellent way to become independent of public sources, which often do not take enough care about the average customer.

## Advantages Of An Off-Grid Solar Power System

If you are interested in investing in a solar power system for your residential or commercial property you probably know that there are two broad categories that you can choose to go with; these being grid connected or off-grid systems.

Grid connected systems, as the name implies, are solar power systems which are wired into the main power grid and therefore come with various benefits and downfalls. If you are interested in taking advantage of various remaining financial incentive programs which are still on offer, then this is the option for you, and if you are eligible for any of these solar rebate schemes, you stand to make financial gain from your system on top of the reduced power bills.

Many people are staying in areas where it is very feasible and easy to start living off grid, yet most do not realize it or do not understand their true benefits.

Off-grid power systems appeal to others who want to gain complete independence from the main power grid. This can be the preferred option for various reasons, such as remoteness of location. Many individuals or households who choose to go with an off-grid solar power solution find that it makes financial sense to do so because of the high cost of grid connection in extremely remote locations. Others choose off-grid solar because they feel the urge to be completely independent of the main power grid and drastically reduce their carbon footprint and culpability for global warming and climate change.

It benefits the environment by slowing down the consumption of non renewable sources like fossil fuels.

It is becoming more feasible for people to use off grid power due to the reduced costs that retailers are charging for items like solar panels and wind turbines. Despite this decrease in costs, many people still find them unaffordable as they can still cost $10,000s of dollars. One great discovery that many people like myself have discovered is that these systems can be built DIY for less than $200.

Off-grid solar power systems can initially involve higher start-up costs due to more panels being necessary as well as several high-capacity, deep-cycle batteries which are used to store excess power during high yield periods during the day for use when solar energy is not available later in the night.

The internet has allowed more people to gain access and knowledge to homemade energy systems and their benefits. They now know how traditional methods of using fossil fuels are damaging the environment permanently and the amount of money they can save by using off grid power systems.

There are certainly some downfalls involved with off-grid solar; these involve the complete dependence that is placed on the system. With a grid connected system, if you use more power than your system has generated through the day you simply begin to tap into the grid's power supply, whereas with an off grid system if you happen to use more power than your system can generate you do not have any options left. This means that more forethought is necessary before investing in an off-grid system, you should consult with an electrician to find a system which best suits your needs, and during times of low power yields or cloudy weather increased energy efficiency measures and rationing of power may be necessary.

**Equipment For Off-Grid Solar Systems**

- Solar Charge Controller
- Battery Bank
- Off-Grid Inverter
- Solar panels
- Inverter
- Back-up generator
- DC disconnect
- Batteries
- AC breaker panel
- Kilowatt-hour meter
- Utility disconnect
- Rectifier
- Electrical wiring

# HYBRID SOLAR SYSTEMS

A hybrid solar system typically uses two solar systems that work together to increase efficiency. A typical (PV) Photovoltaic system (electric) is only 15% efficient. By comparison a car engine is about 35% (efficient - a typical car engine produces heat which is lost.) A PV system is usually designed to produce around 80% of a home's electricity. With a hybrid system these two usually independent technologies work together so that efficiency is increased and now more than 80% of a household's energy needs is produced. By combining two systems through a computer, the overall efficiency of 50% can be achieved- and noted this is almost double than use of a single system. However, incorporating two usually independent systems is not inexpensive. A hybrid solar system does need to be designed to work together. There has to be some sort of control mechanism which sends the power where it is needed.

Typically the two types of Solar Systems that are incorporated into a hybrid unit are PV Electric and Hot air. A PV system is installed on the roof which provided electricity for the home or business. Since the Solar Panels are usually raised off of the roof-companies use this space to heat the air under the panels. As the air warms, it heats and rises up the roof. Near the top of the roof the air is ducted inside the attic where it can be used for a variety of purposes. Once in the attic hot air can be ducted into the building rooms directly for heat.

This is very efficient as long as the outside air is cooler than air in the building. So in winter the hot air can offset the overall energy usage. This concept can work well in cooler climates. In the summer months, the system can work in reverse. If the climate is hot the cool night air can be ducted into the building as the air flowing under the panels at night is cooled by radiant heat loss.This is similar to an air conditioning system or even a whole house fan. There may be limitations to using a solar air heating systems though as certain extreme climate regions may reduce their efficiency. In areas of high humidity- direct air systems will probably not work well. However, this does not mean one should not consider this option.

The hot air can also be used to heat water, however- heating water from air is highly inefficient. It is more efficient to heat water near the solar PV panels, which could then be used to heat air through a heat exchanger. Hot water can be stored and used when needed for a variety of uses. Hot water is a useful commodity for daily use, including showers and cleaning. But hot water can easily be used to heat the air in the building as well. Hot water can also be stored in tanks and used when needed. So perhaps an integrated hybrid system is the next wave of the future. It will be interesting to note- if the hybrid systems use air or water to gain maximum efficiency.

## Advantages Of Hybrid Solar

Installing a solar energy system for your home is much easier than you think. Having such a system will turn your home into a green power plant that is harnessing the sun's energy and converting it into the electricity you need for your home. Your current electricity supply from your utility company is most likely based on a coal-powered plant that burns fossil fuels, emitting carbon dioxide and air pollution. There are a lot of benefits of going solar and generating electrical power from the sun.

***Its a huge benefit to the planet.*** Utility companies produce the highest percentage of electrical energy and they do this by burning fossil fuels and thus releasing carbon emissions into the atmosphere. Switching your electrical supply to a solar energy system is one of the biggest contributions you can make to saving our planet and counteracting global warming.The pollutants emitted from the utility company is one of the biggest contributor to the green house effect. Cutting down on the electricity we draw from the grid will mean less carbon emissions resulting into a cleaner environment for us and the next generation.

**_Huge Savings._** Most homes will only achieve being a hybrid solar energy home. Every kilowatt (KW) of energy that you produce with your solar panels represents a KW of energy that you do not have to pay for from your utility company. This savings could be as much as 60% to 80% of your utility bill. Imagine what that could mean to your family. That is more money for groceries, school supplies, vacation and much more

It increases the value of your home. Going solar is one of the upgrades that you can make to your home that will guarantee to increase the value of your home. Because the result of the upgrade is measurable and can be proven by the reduction of your utility bill. Even if you sell your home, it will appraise for higher than a home without the solar panels. And its up to you, you could either sell your home for the increased value it has with the solar panels, or you could take your panels with you to your new home. In either case, you win.

## Equipment For Hybrid Solar Systems

- Charge Controller
- Battery Bank
- DC Disconnect
- Battery-Based Grid-Tied Inverter
- Power Meter
- Wires
- Conduit
- A grounding circuit
- Fuses
- Outlets
- Metal structures for supporting the PV modules.

## SOLAR ENERGY

Do you know that the global demand for solar energy exceeds than its supply? It has several uses through its photovoltaic cells for water heating and production of electricity and even drying our clothes. The energy also generates lighting for both outdoors and indoors. It could even be utilized to power cars, for small appliances and calculators and watches. Also, it is also widely used to heat swimming pools. The best thing regarding these facts about solar energy? You can even cook your food with it. That is why it is fair enough to say that solar power as compared to some traditional forms of energy, is better for our environment.

## SOME FACTS ABOUT SOLAR ENERGY

- Solar energy is a renewable resource
- A home solar system is usually comprised of solar panels, a battery, a charge controller, inverter, wiring and support structure.
- Solar panels come in a variety of colors to choose from.
- The system typically has a 5-year warranty, though the solar panels are guaranteed for 20 years.
- Solar energy could be gathered and kept in batteries, insulated, absorbed, reflected, and transferred.
- A 1 KW home solar system comprises of almost 10-12 solar panels and needs approximately 100 square feet base area.
- Do you know how long it takes to set up a 1-KW home solar system and how much you would likely to spend on it? It may only take up to 48 hours and it may cost you US$ 10,000. However, it could actually differ significantly and does not take into consideration any incentives presented by the state.

- Depending on the battery back up, a solar panel system could give electricity 24 hours a day, 7 days a week, even at night time and even on cloudy days.

- How about knowing some sun-related facts about solar energy?

- Pollution, clouds and the wind could actually block the sunlight's rays to reach our planet Earth.

- Our sun is the primary source and supplier of fossil fuels (non-renewable) such as petroleum, coal and gas, that started existence as animals or plants whose energy originated from the sun millions of ages ago.

- Solar energy is likely the one causing currents in the oceans and weather and climate patterns.

- Shell Oil foresees that by the year 2040, 50% of the global energy would come from renewable sources.

- Did you know that the appliance which consumes the most amount of electricity is the electric oven? Only followed by microwaves and central air conditioning. Do you think it is about time you use your electric ovens more sparingly now?

- In only 1 hour, more sunlight falls on our planet than what is consumed by the whole population in a year.
- Calculating for just about 5% of the global population, Americans use 26% of the world's energy.
- Nearly 2 billion people on earth are presently without electricity. So you can just imagine how lucky enough that you are not one of them. Solar energy can be used to heat water, dry clothes, heat swimming pools, power attic fans, power small appliances, produce light for both indoors and outdoors, and even to power cars, among other things.
- You can install solar energy in remote locations.
- If there is a power outage, you can still have electricity!
- As your energy needs grow, you can add more solar panels
- Solar panels operate silently.

- Solar powered cars are not yet up to par with other cars. They have a much slower speed.
- Solar energy systems require very little maintenance and will last for a long time.
- Technology for solar energy is constantly improving.
- To run a solar energy system, you do not even need to connect to a gas or power grid.
- Solar energy definitely has its benefits like reducing your home energy costs and not to mention caring for the environment. Obtaining usable energy from the sunlight has been used for centuries by modern technologies and there's no reason why we should not give it a try.
- It is easy to set up, plus
- It is not expensive.
- It is a great way to illuminate your backyard or your garden. You need a relatively large area to install solar panels if you want a good level of efficiency.

# THE FUTURE OF SOLAR ENERGY

With the current concentration on global warming, the existence of which is denied by only a few naysayers, developments in solar power are moving at a pace to fill the need for efficient green solutions to our seemingly insatiable desire for energy.

There are many people who are in favor of solar energy. However, the cost of this has not been very effective in recent years. Today, technological innovation is taking solar energy prices to competitive levels downwards. It is leveling with other renewable energy sources.

The idea of generating unlimited energy and free of contaminants from the sun is interesting.

Unfortunately, with oil prices falling in recent months, it affected the progress of solar energy. Generating electricity with solar energy costs between 25 and 50 cents per kWh, which is between 5 and 8 times more than it costs to do with fossil fuels.

The price has been and remains an obstacle to renewable energies dominate the energy industry, so it should continue working on ways to improve the efficiency of small solar panels and reduce production costs, so that in the next years we can be on par with fossil fuels.

The companies are already working on methods to simplify the construction of solar cells and cells that use a fraction of the material of the former, both works contribute to reduced production costs of solar panels and therefore in the final price.

Jobs in Nanotechnology are also being made to reduce the amount of material used and to improve the efficiency of solar panels, it helps to convert into electricity more sunlight than they receive.

This last decade has made great strides in lowering costs of solar energy, but we must keep working, as even that is not as cheap as fossil fuel power generation are not able to see a takeoff on the use of solar technology.

The future of solar energy is related to events that today they live, the production of solar devices has grown 30% in recent months and this is because there are more people who see PV as a viable solution to replacing fossil fuels. If this continues to happen, after 10 years, with a population much more awareness, the production of solar devices will be greater because they will have more demand.

# THE IMPORTANCE OF SOLAR ENERGY

### ✦ *Cut-down electrical bill*

The key reason that most households convert their power source to solar energy is to cut down their electrical bill because the electrical usage generated from the sun is free. By converting as many home appliances as possible to use solar energy, you can save a significant lump-sum of money uses to pay for your typical electrical bill. It is the best option to reduce utilities expenses.

### ✦ *It is a renewable energy source*

Typical electricity is generated from fossil fuel that will run out one day. Solar energy is a good alternative to replace fossil fuel as the major energy source because solar power is renewable at absolutely no cost to supply energy infinitely.

### ⊹ Environment friendly

The world pollution is getting worse. Any effort that can reduce the pollution to the environment helps to save the earth. Solar panels are able to harness the energy from the sun and convert it to electricity. Therefore, the use of solar panels is environment friendly. Therefore, solar energy that is harmless to the environment will be the major energy source for future starting today.

### ⊹ Low or no maintenance needed

Once you have installed the solar power system, it can last twenty to thirty years without major maintenance needed. You may need to do system check once a year, just to make sure everything is performing as it should. Since it requires very minimum maintenance cost, your cost should be minimal.

# BASICS OF ELECTRICITY FOR SOLAR POWER INSTALLATION

You may remember electrons from high school science class. They are negatively charged particles that orbit atomic nuclei and are so tiny that 166 trillion of them could fit on the point of a sharp pencil. These electrons hold matter together because their charge is the opposite of the positive nuclei. However, if an external charge from say a battery is applied to a conductive material such as copper, electrons will start moving through the conductor. When they meet resistance such as in a light bulb or a motor, they will start to do some work and we have light or power. Volts, Amps, and Watts. The Big Mystery.

In order to understand how solar power works and how to correctly install a system, you need to see the relationship between these three concepts. Voltage is a measure of electrical potential. You can think of it as how much motivation the electrons have to move along the conductor. If voltage is high, they really want to move and can even jump a gap in the conductor such as in a car's spark plug. Amperage represents the amount of current that passes a given point in a given time frame. Wattage is the amount of electrical energy consumed or the amount of work done by the electrical circuit. These terms are defined mathematically in the equation:

*Watts = Volts x Amps*

*or Volts = Watts / Amps or Amps = Watts / Volts*

These equations will allow you to make all the calculations that you need to plan your solar system. For example, you will be able to calculate the amount of watts required for a space heater that draws 4 1/2 amps. Watts = 220volts x 4.5amps = 990watts. How is this relationship applies to solar. Solar panels are usually wired together in an array of 5-10 panels and one of the key questions is how to connect them. Copper wire is used to connect the panels, and because of the resistance of the wire, there is a loss of current which increases with the length of the wire. The key is voltage since a thin wire can carry a high voltage but a lower amps. For example, your auto's spark plug wires will carry 30,000 volts which easily jumps the plug gap, but the amperage is negligible. If you have 5 12volt, 135 watt solar panel connected and you want to connect to a battery 25 feet away you would need a cable that carries:

$$amps = (135 / 12) \times 6 = 67.5 \text{ amps}$$

Checking a wire size chart this would require a 4/0 cable which is 12mm or almost 1/2 inch in diameter. At the cost of copper these days, that will be a small fortune, not to mention the difficulty of installation. If instead, the panels are wired so they output 48volts, your amperage would be: amps = (135/48) x 6 = 16.9 amps and a wire would be required which is only

3.3mm or about 1/8 inch in diameter. This is a much more manageable and cost effective solution.

## What are Kilowatts?

In order to size a solar system, one needs to calculate the Watt Hours required. This formula is Watt Hours = Watts x Hours Used. For example, the space heater above is used for 2 hours per day so: Watt Hours = 990watts x 2hours = 1980watthours (Often referred to as Watts) You would have to figure the usage for all appliances in your house to come up with the total watt hours required. You also could look at your electric bill if you are on the grid and it will show you your usage often expressed as kilowatts.

1000 watts = one kilowatt (kW)

1000 kilowatts = one megawatt (MW)

1000 megawatts = one gigawatt (GW)

# ELECTRIC CIRCUIT

An electric circuit is formed when the electrons from a voltage or current source flow, but most circuits have more than one device that receives electric energy. Most of the devices in a circuit like a light bulb, resistor, or a capacitor are connected in one of two ways, series or parallel. When it is connected in series, the devices form a single pathway for electron flow between the terminals. Then when it is connected in parallel the wires form branches; this means that it separates the path for the flow of electrons. Parallel and series both have their own different way to connect and they are calculated using different formulas.

Then there are different examples of series and parallel circuits.For example in earlier automobiles both headlights went out when one bulb burned out. The headlights must have been connected in series because if one of the headlights burned it would cut the electric flow to the other bulb as well, making them both out of operation. Examples of parallel circuits are used all over homes like multi-bulb lights and Christmas lights.The parallel circuits make sure that if one of the bulbs burn out then the others will still light up.

### Series Circuit

Series is an electric circuit in which electrical devices are connected along a single wire so that the same electric current flows through all of them. What this means for resistance is that it is greater, because the electrons all go through the same path through the circuit. The way to find series is to find the total resistance by using the formula: $Rt= R_1+ R_2+ R_3$. This means by using the different amount of resistance in the circuit then adding them together, so that the total resistance is calculated. Then since the electrons go through in one path then the current goes slower. In contrast to a series circuit, the electrons in a parallel circuit do not go through just one loop in one pathway. The electrons go through different loops so this means the resistance is not greater compared to series; the current is greater because they have more pathways to go through.

## Parallel Circuit

Parallel circuit is an electric circuit in which electrical devices are connected in such a way that the same voltage acts across each branch, and any single one completes the circuit independently of all the others. We can calculate the total resistance in a parallel circuit by using the formula: 1 over Rt = 1 over R1+ 1 over R2 + 1 over R3. By using the formula the total resistance is found, but an important thing to keep in mind is to reverse the answer.

## Ohm's Law

Probably the most important mathematical relationship between voltage, current and resistance in electricity is described as "Ohm's Law". In 1827, George Ohm developed his well-known formula concerning electricity after performing various experiments and studies. Ohm's formula is used to find out the required resistance, voltage or current values so that we can design circuits and choose the right components. For example Ohm's law is used to determine the correct resistor value in a circuit when the voltage is known and you would like to limit the current to a certain value.

Ohm's Law is defined as $V = I \times R$, whereby V is the voltage, I is the current and R is the resistance (in Ohms). When using the equation in practice, the value of all of the components can be more easily determined by rewriting the equation. When you would like to find the current you can use $I = V / R$ or when you like to find the resistance value you can use $R = V / I$.

If we write Ohm's law as $I = V / R$, it let us know that the electrical current in a circuit can be calculated by dividing the voltage by the resistance. In other words, the current is directly proportional to the voltage and inversely proportional to the resistance. And so, an increase in the voltage will increase the current provided that the resistance is held constant. Alternatively, if the resistance in a circuit increases and the voltage does not change, the current will decrease.

If you will want to determine the voltage in the event that the resistance as well as current are known, you can utilize the formulation $V = I \times R$. The formula shows us that if either the current or the resistance increase in a circuit (when the other stays the same), the voltage will also have to increase.

The resistance in a circuit may be computed with $R = V / I$. When the current is kept constant, a rise in voltage will result in a rise in resistance. An increased current while voltage stays constant will lower resistance. It must be noted that for a wide variety of materials used as a resistor (such as metals) the resistance is fixed and does not depend on the amount of current or voltage. In semi-conductors however, the resistance is often dependent on the current or voltage level.

To get a better understanding on the mathematical relationship between voltage, resistance and current, Ohm's Law is very useful.

## VOLTMETER

A voltmeter is an instrument that is used to check and measure the amount of voltage that is passing between two points in an electric current. It measures the amount of positive electrical charge as it enters one point within an electric circuit and then measures the negative input as it passes through another point.

### *How a voltmeter works*

A voltmeter in simple terms comprises of a red positive terminal, as well as a black negative terminal and a display. Most types of multimeter include a voltmeter function.

A good case example to demonstrate how it is used is when trying to determine the amount of voltage left in a battery. Two wires are used; one wire is connected from the positive terminal of the voltmeter, to the positive end of the battery. The other wire is connected from the negative terminal on the voltmeter to the negative point on the battery.

It is important to note that mixing up the positive and negative connections, for instance connecting the positive terminal to the negative end of the battery can severely damage the voltmeter, particularly so if it is a needle point voltmeter.

*Types of Voltmeters*

A **needlepoint** voltmeter as the name suggests simply points out the number representing the voltage amount, just the way a watch with a minute hand pointing out the number of minutes into the hour. The disadvantage associated with this type of voltmeter is that at times the voltage being measured may be too weak to be able to push the needle to the correct reading, and ends up giving too low a voltage value.

The digital voltmeter therefore will indicate the voltage amount in a manner similar to the digits displayed by a calculator. These voltmeters are usually calibrated to display higher accuracy readings than their counterparts.

## Handling Voltmeters

Voltmeters, and multitesters generally, tend to be fragile instruments that can be damaged by thoughtless actions like using a voltmeter made to measure small amounts of current to measure a large voltage. However, another unwise move is to use a voltmeter intended for measuring large voltages to measure small amounts of voltage. In this scenario, although the voltmeter will not get damaged the readings one will get will not be accurate. Both extremes should be avoided.

Another precaution to take is to first determine whether the voltage being measure is either a direct current, that flows in one direction, or is an alternating current, that flows back and forth. Different voltmeters are made to handle these different electric charges. Even in cases where, a single is capable of measuring both kinds of voltages, it must first be prepared as to whichever current it is to measure without which the voltmeter will be damaged to display faulty voltage readings.

One final caution is that while using a voltmeter, one should take care not to touch any of the terminals with their bare fingers as this will most likely result in being electrocuted, at times even to death. As much as possible, measuring of a voltage should be a hands free maneuver with the aid of alligator clips.

## AC TO DC POWER CONVERTER

There are different ways to convert Alternating Current (AC) into Direct Current (DC). Each device requires different types of power supplies and therefore it is quite essential to identify the precise type of current required for each device before choosing an AC to DC converter.

### Some Factors to Consider

**Maximum Load Required:** Before choosing these converters, it is essential to consider the maximum load the circuit requires so as to ensure optimum functionality. Most of the electrical devices include load ratings which must be taken into consideration while fixing converters.

**Right Power Supply Output:** There are different types of power supply output such as regulated, linear or unfiltered and filtered power supply. The filtered power supply units are considered to be an ideal source as it facilitates in reducing the frequency noise that can emancipate from a power device.

**Watt:** Another factor to consider before buying converter is its wattage. It is essential to check the watt of the electrical devices before choosing power converters.

## Working Procedures

The AC to DC converter utilizes an electronic component known as a rectifier to convert the power supply. There are different types of rectifiers such as the single diode rectifiers and the bridge rectifiers. The basic function of a rectifier is to accept an electrical output and to direct the flow of current into a single direction. This process creates direct current. All rectifiers are designed to accept a particular level of current and therefore it is essential to choose the right adaptors before it is used with AC power supply. Using the wrong type of rectifiers can cause irreparable damages to the electronic devices and converters.

Different electrical equipment requires different level of direct current. Majority of the users prefer to use adaptors with higher amperage as these devices are quite beneficial. These devices draw the right amount of power from the source which is safe for the electrical components to use.

The Alternating Current to Direct Current converters includes various components. However, the main components include the wattage output and the wave output. The converter converts both sine waves and square waves quite easily. Although both these waves are quite similar in features, there are minor differences which affect the functions of sophisticated electrical equipment.

Today there are various types of electrical devices which can be bought from online stores at affordable price rates. The AC to DC converter is among the most widely used electrical device as it is quite versatile and useful at homes and commercial spaces. It is known to be one of the most convenient ways to convert electric power effortlessly.

As there are various reputed manufacturers and sellers listed in various online stores, it is quite essential to choose the right type of device from the right dealer. There are many online sites which offer electrical devices at discounted price rates too. Most of the online stores showcase converters with different attractive features which include different watt and power features.

## EFFECT OF LOAD RESISTANCE ON POWER

Load resistance always added in the overall resistance of the circuit. Which has so many effects on the circuit such as with increase in load resistance, the voltage drop will be increased due to that there will be sudden increase in the heat, which can burn the insulation of the circuit.

## EFFECT OF TEMPERATURE ON POWER

Photovoltaic solar panels convert sunlight into electricity, so you would think that the more sunlight, the better. That is not always true, because sunlight consists not only of the light that you see, but also of invisible infrared radiation, which carries heat. Your solar panel will perform great if it gets a lot of light, but as it gets hotter, its performance degrades.

# EFFECT OF INSOLATION ON POWER

Insolation is the incident solar radiation onto some object.Not all of the solar energy that reaches the earth actually reaches the surface of the earth. Although 1367 W/m2 of sunlight strikes the outer atmosphere, about 30% of it is reflected back into space.

## RADIATION POTENTIAL

Radiation is something many people do not know much about. Radiation is in fact everywhere, you just can not see it - well, most of it that is, and the kind you can see, you can not see all that well. Most people have a general idea of what it might be, and might even know a few examples of how it manifests itself - such as from a microwave or x-ray.

With the recent threat of nuclear disaster in Asia continent, sparked from the damage inflicted by the massive earthquake and subsequent tsunami, many people have no doubt been caught not knowing as much as they would have liked about nuclear radiation effects, types of radiation, and what a "safe" radiation dose amounts to.

But before we can understand what makes radiation so potentially harmful, we have to understand what it is and where it comes from. There are two main types of radiation: ionizing radiation and non-ionizing radiation. Ionizing radiation is what we are looking for here, and it is the kind of radiation that turns atoms into ions, or atoms with uneven amounts of protons and electrons.

### "So What Exactly is Ionizing Radiation?"

Ionizing radiation is simply radiation that has the energy-capacity to ionize atoms, and is often the sole form of radiation implied when speaking of radiation. The ionizing of an atom occurs when ionizing radiation collides with an atom, "knocking out" an electron and causing an uneven amount of electrons and protons. This leaves the atom with a net positive charge - also called a cation.

Conversely, a net negative charge occurs when an atom gains an electron due to a free electron that is energetic enough to literally force its way into an atom, also called an anion. These two processes are the basis of ionization. Alpha particles, beta particles, neutrons, x-rays, gamma rays, and cosmic rays are all examples of ionizing radiation.

Now that we understand the process of ionization and what ionizing radiation is, we can uncover its potential dangers. As simple as this may sound, the adverse physical effects of radiation are caused by the alteration of atoms by this ionization process to the point of manifesting physical symptoms, such as cell death, genetic mutations, cancer, and ultimately, even death.

*"Where does this ionizing radiation come from?"*

Think of ionizing radiation as invisible particles or waves of energy that are emitted from either radioactive atoms or radiation-producing machines such as nuclear reactors. Radioactive atoms, also called radioisotopes or radionuclides, are atoms with an unstable nucleus and are therefore experiencing radioactive decay at a rate expressed by its half-life.

During radioactive decay, the atom emits ionizing radiation in the form of gamma rays and/or subatomic particles. However, the amount of ionizing radiation emitted from most naturally occurring radioactive decay is within safe limits. Nuclear reactors on the other hand, are responsible for the perpetual emission of large amounts of ionizing radiation through nuclear fission.

Of course, this radiation is contained within the housing structure so long as it does what it is supposed to.Which is exactly the point in question! If something happens that is outside the control and foresight of engineers, such as earthquakes, and other natural disasters, and the radiation is somehow allowed to leak out or God forbid, flood out there will be a major disaster.

The ionizing radiation in the form of radioisotopes such as iodine-131 and caesium -137 will then be dispersed by the wind which will carry them far and wide. The lifespan of a radioisotope is determined by its half-life, therefore you can say that the amount of damage it can inflict is partially based on its half-life. Iodine-131 for example, only has a half-life of about 8 days, whereas caesium-137 has a half-life of about 30 years.

But half-life is not the only factor involved in determining potential danger and its extent. As the half-life decreases, the amount of ionizing radiation emitted per unit of time increases. So although the time during which it is doing damage is shorter, it is also more concentrated and intense. Another factor is the atomic mass of the radioisotope.

The heavier it is the quicker it will sink to the ground, and the smaller the radius of contamination. Conversely, the "lighter" the radioisotope, the further it will be carried by the wind. Of course, penetration into the ground water and any nearby bodies of water such as rivers can be particularly dangerous due to the range of contamination and potential imbibing of radioactive water.

## ROOF PERFORMANCE FOR SOLARS

Have you ever wondered whether your house roof is suitable for a solar photovoltaic installation? Follow a few rule of thumb points and get a better idea.

In domestic situations, the roof is usually the most appropriate location to install solar photovoltaic panels. This is because the roof slope is at a convenient angle and high enough to avoid serious overshadowing problems.

### Roof Orientation

The main feature that will determine your house's suitability for solar PV is the orientation of the roof slope. Solar PV panels are best installed facing due south, but any orientation on the southern quadrant of the compass can be used, even as far as the south-east or south-west. At these extremes, the electricity generating capacity will be reduced by a small, but acceptable margin.

### Which Way is South?

If you have a copy of the house plans, there will usually be a north point from which you can easily establish south. If you do not have drawings, you will need to locate the direction of south with a compass.

## Is the Roof Overshadowed?

It is important to ensure there are no obstructions that will block the sunlight from reaching the solar panels. Even a small amount of shading can reduce the performance of the whole system, not just that of an individual shaded module. Looking from the same height as your proposed solar installation, and working from east to west, check there are no obstacles such as trees or buildings that can obscure the sun at its lowest winter height.

## Identify Obstacles

You should take into account the future growth of trees and shrubs, as vegetation may shade the system after only a few years. Look for other obstructions, as cables and aerials attached to the building can also cast a problem shadow over the proceedings.

### Available Sunlight

In the UK, the amount of available sunlight varies with the geographical location of the site. For instance, there is more available light in the West Country than there is in Scotland. Furthermore, the light energy varies during the year with more daylight available in the summer and less in the winter.

### Is Your Roof Pitch Suitable?

The high latitude of the UK means that the optimum roof angle for solar PV panels to achieve maximum performance is about 45 degrees. However, most domestic roofs have a shallower pitch than this, set at between 25 and 35 degrees tilt. The additional energy that would be gained by increasing the pitch of the array out of the plane of the roof would not normally be justified on either cost or appearance grounds. As a result, it is recommended that the solar PV array be installed at the existing pitch, provided this is not less than 25 degrees and not more than 60 degrees.

### Is Your Roof Strong Enough?

Individual solar PV panels are not usually heavy, but when multiple panels are combined the weight can become significant. You will need to be satisfied that the roof structure is strong enough to carry the solar PV installation. You may need to arrange for an inspection by a local qualified building professional to establish this.

## Storing the Photovoltaic Equipment

You will need to identify a suitable location to store the electrical equipment that comes with the PV installation. Ideally, the cables should be located within the structure of the pitched roof. The inverter and other electrical equipment can be placed in the roof void, if it is not being used. As far as possible, it is important to position the electrical equipment close together in order to keep the cable lengths as short as possible.

## Solar PV Installers

When you are satisfied that your house is a contender for a solar PV installation you can then approach certified solar PV installers and obtain more detailed information and cost estimates for the work. PV installers will have access to specialist instruments and sophisticated software to provide you with accurate information about your proposed solar installation.

## SOLAR SYSTEM SIZING

### ⊥ Battery Sizing

In order to size your battery, you need to double your initial Watt-Hours value in order to make it so your loads only drain the battery down to 50%. You will take that last wattage value you calculated and multiply it by 2. You then divide it by the voltage, either 12V, 24V, or 48V based on what controller you end up using to find the Amp-Hours needed.

## ⊥ *Inverter Sizing*

For home solar goers, the most important thing is to install a system that performs very well and helps reduce the entire electricity bill. Thus, optimizing the power generating system is by all means necessary, and sizing the solar inverter that works for the system is one of the best choices, for a solar inverter works to convert the direct current from solar panels into alternating current, which is used to power all the appliances of a house.

Before deciding to calculate the size of the inverter for the system, you have two important things to consider: the total watts of all the appliances, which indicate the amount of electricity required by the setup, and the total units because they decide how much electricity can be used each day.

Besides this, today's home solar systems can be generally divided into three types: standalone, grid-tie and battery backup systems. Different systems require different types of solar inverter installations.

Sizing an inverter for a solar power system is closely connected to the total watts of the family's total appliances. The input rating of the converting device you use cannot be lower than the entire watts of these appliances. If you have fridges or pumps connected, you need to take serious notice of this advice. At the same time the size of the device should also match the watt of the solar panels installed on your rooftop. If it is in a standalone system, where batteries are used for the sake of storing direct current for re□uired use, its nominal input voltage should be the same as that of the batteries.

Although you have learned the basic knowledge of sizing an inverter, you still need to be clear of the effects of under-sizing and over-sizing, because people who wish to go solar may meet these problems, which will influence the ultimate performance of your solar power system.

Simply put, a too big inverter can cost more of your budget, while a too small one cannot meet your required amount. In detail, the alternating current output is decided by the converting device not by the output direct current amount from the solar panels. If the solar inverter is not big enough, some direct current will be wasted in the converting process. On the other hand under-sizing an inverter can overheat the parts used in the system and make their lifespans become short. If it is over-sized, the whole system's efficiency will be decided by the average point of operation. Therefore, its efficiency will shrink according to how much you have over-sized your device. Also, if you wish to add more solar panels, the system's total performance will discount too.

Apart from sizing solar inverters, sizing solar panels and batteries also needs deep concerns, and this can make your system work better too. But these knowledge and experience cannot be absorbed in one day, and they need a long time practice and careful calculation.

To size the inverter you need to add up all the wattages of all the items you want to run. You then need to pick an inverter with more wattage than this. Also, make sure your inverter matches your battery bank voltage as well.

### ⊥ MPPT Solar charge controller Sizing

Next, you need to find a controller that can accept the wattage you need. You can check the controller specification sheet to see the wattages they can handle. For example, a 30 Amp Controller can handle 400W on 12V, so you know you can have up to 400 Watts on there.

MPPT solar controllers are very clever and extremely efficient. They are relatively expensive .They like higher voltages.

# TRANSFORMER PROTECTION

Transformers of varied sizes and configurations are at the heart of all power systems. As a critical and an expensive component of the power systems, transformers play an important role in power delivery and the integrity of the power system network as a whole. Transformers, however, have operating limits beyond which the transformer loss of life can occur. If subjected to adverse conditions there can be a heavy damage to the system and system equipment, besides intolerable interruption of service to the customers. Since the lead time for repair and replacement of transformers is usually very long, limiting the damage to faulted transformers is the foremost objective of transformer protection.

## *Economic impact of a transformer failure*

- The direct economic impact of repairing or replacing the transformer.

- The indirect economic impact due to production loss.

Operating conditions like transformer overload, through faults, etc often result in transformer failure, highlighting a need for transformer protection functions, such as over excitation protection and temperature-based protection. Extended functioning of the transformer under abnormal condition such as faults or overloads can compromise the life of the transformer.Adequate protection should be provided for quicker isolation of the transformer under such conditions. The type of protection used should reduce the disconnection time for faults within the transformer and minimize the risk of catastrophic breakdown to simplify eventual repair.

### Transformer Failure

The risk of a transformer failure is two-dimensional: the frequency of failure, and the severity of failure.

Most often transformer failures are a result of "insulation failure". This category includes inadequate or defective installation, insulation deterioration, and short circuits, as opposed to exterior surges such as lightning and line faults.

## Failures in transformers can be classified into

- Winding failures resulting from short circuits (turn-turn faults, phase-phase faults, phase-ground, open winding)
- Core faults (core insulation failure, shorted laminations)
- Terminal failures (open leads, loose connections, short circuits)
- On-load tap changer failures (mechanical, electrical, short circuit, overheating)
- Abnormal operating conditions (overfluxing, overloading, overvoltage)
- External faults

## Other causes of transformer failure may include

***Overloading*** - Transformers that experience a sustained loading that exceeds the nameplate capacity often face failure due to overloading.

***Line Surge*** - Failure caused by switching surges, voltage spikes, line faults or flashovers, and other T and D abnormalities suggests that more attention should be given to surge protection, or the adequacy of coil clamping and short circuit strength.

***Loose Connections*** - Loose connections, improper mating of dissimilar metals, improper torquing of bolted connections etc can also lead to failures in transformers.

***Oil Contamination*** - Oil contamination resulting in sludging, carbon tracking and humidity in the oil can often result in transformer failure.

***Design or Manufacturing Errors*** - This includes conditions such as: loose or unsupported leads, loose blocking, poor brazing, inadequate core insulation, inferior short circuit strength, and foreign objects left in the tank.

***Improper Maintenance or Operation*** - Inadequate or improper maintenance and operation are a major cause of transformer failures. It includes disconnected or improperly set controls, loss of coolant, accumulation of dirt and oil, and corrosion.

***External Factors*** - Several external factors like floods, fire explosions, lightening and moisture can be established as the causes of the failure as well.

## Transformers Protection Best Practices

Transformer failures and safety hazards can be avoided or minimized by ensuring that the conductors and equipment are properly sized, protected and adequately grounded. Incorrect installation of transformers can result in fires from improper protection, as well as electric shock from inadequate grounding.

- Once the transformer is placed, the tank must be permanently grounded with a correctly sized and properly installed permanent ground.
- Access should be restricted to the transformer liquid-filled compartment in conditions of excessive humidity or rain.
- Dry air should be continuously pumped into the gas space if humidity exceeds 70%.
- Transformer should be given protection against rain such that no water gets inside.
- All equipment used in the handling of the fluid (hoses, pumps, etc.) should be clean and dry. If

the insulating liquid for inspection is drawn out, its level should not go below the top of windings.

- Sufficient gas pressure must be maintained to allow a positive pressure of 1 psi to 2 psi at all times (even at low ambient temperature) when liquid-filled transformers are stored outside.

- Final inspection of the transformer is essential before it is energized. All electrical connections, bushings, draw lead connections should be checked.

- Upon loading the transformer should be kept under observation during the first few hours of operation. All temperatures and pressures should be checked in the transformer tank during the first week of operation.

- Surge arresters must be installed and connected to the transformer bushing or terminals with the shortest possible leads to protect the equipment from line switching surges and lightning.

# SURGE PROTECTION DEVICE

There are many parameters that have to be met when one is selecting that ideal Surge Protection Devices as well as circuit breakers. There are many devices that can be used, but it is always important to assess the risk that is involved.

**Rules to be considered**

It is important to appreciate the fact that Surge Protection Devices are very important. You need to select the best, so as to ensure that you are well protected at all times.

*1. Get familiar with the available options.* There are many types and categories that you can think about. This is the only way in which you can make the most suitable selection for your needs.

*2. You also have to assess the risk of lightning striking as well as the discharge capacities.*

**3. Protect the Surge Protection Device.** It is very important to protect it as well. This makes everything much safer.

## Devices

There are three different types of devices. To make a distribution switchboard safe, then the Type 2 SPD shield is adequate. You, however, need to concentrate on the discharge capacity.

## Risk assessment

It is usually very complex to assess the risk involved. It is a very painstaking process as well. The most important thing is to think of areas where there may be a lot of risk and areas that are not that risky. After this, it should be easy to choose the best SPD that is the best for you. You should evaluate the building that needs protection.

Lightning is a common occurrence in many parts of the earth. There are areas that are high risk than others. Some areas do not experience any lightning at all. The risk with lightning is not the density. Different areas have their own assessment standards. There are countries that make the use of standards a compulsory thing when one is considering a Surge Protection Device for the big buildings that are very sensitive like data centers, hospitals, and industrial facilities.

To be safe, you should always go for the Surge Protection Device. If the area between the Surge Protection Device and the protected equipment happens to be more than 10 meters, then the Surge Protection Devices should be installed on both sides.

## Protect the Surge Protection Devices

Usually, the Surge Protection Devices do not trip. However, there are many scenarios that may happen and they include:

Thermal runaway that may be caused by some currents when it does not exceed the lightning attributes leading to the destruction of the internal components even though at a slow pace.

Short circuit because of exceeding the flow capacity when at maximum. It could also be because of faults that are under Hz from the distribution network such as phase-neutral inversion, neutral rapture and so on. The disconnection is usually provided by one that uses a short circuit, which can be either integrated or external. The device can be a circuit or fuse breaker.

Sometimes it may be necessary to go for an external circuit breaker. However, today many of the manufacturers have them incorporated into one enclosure. A circuit breaker is chosen according to the current within a building where there is a Surge Protection Device. Residential and commercial buildings use different kinds of breakers.

# CONCLUSION

Solar energy is definitely the future trend of energy. Nowadays, many households have converted their home to be powered by solar power system to take advantage of free and renewable energy from the sun. There are both advantages and disadvantages of using solar power. But, if you can benefit from its advantages and overcome its disadvantages, solar energy is a good alternative for existing fossil fuel energy.

www.ingramcontent.com/pod-product-compliance
Lightning Source LLC
Chambersburg PA
CBHW051313220526
45468CB00004B/1331